R for Quantitative Chemistry

R for Quantitative Chemistry is an exploration of how the R language can be applied to a wide variety of problems in what is typically termed "Quantitative Chemistry" or sometimes "Analytical Chemistry". Topics include: basic statistics, spectroscopic data, acid base equilibria and titrations, binding curves (of great current interest for biomedical applications), Fourier Transforms, and chemical kinetics and enzyme kinetics. An innovative feature is the discussion (as an alternative to the less stable nls packages) of the simplex adaptation subplex (R package) coupled with Monte Carlo analysis to determine confidence intervals for estimated parameters resulting from least squares optimization. Chemists who are interested in learning R as a research tool as well as chemists who are teaching quantitative chemistry, as well as their students will be interested. As most R books approach data analysis from an economic, social, medical, or biological context. Analysis of chemical data draws upon specific numerical models and a different set R programming and packages than is typically discussed in other disciplines. This book will be based upon, in large part, actual experimental data and will include end of chapter questions and projects. Readers are encouraged to email the author at gosserch@gmail.com and to follow the accompanying blog on Medium "R Programming for Quantitative Chemistry".

Key Features:

- Elements of R programming for Chemists
- Literature-based examples
- Includes binding assay analysis
- Integrates theory, experiment, and R programming

Dr. David K. Gosser Jr. is Professor of Chemistry at City College of New York, CUNY. Dr. Gosser received his Ph.D. in Physical Inorganic Chemistry from Brown University.

R for Quantitative Chemistry

David K. Gosser Jr.

CRC Press
Taylor & Francis Group
Boca Raton   London   New York

CRC Press is an imprint of the
Taylor & Francis Group, an **informa** business

A CHAPMAN & HALL BOOK

First edition published 2024
by CRC Press
6000 Broken Sound Parkway NW, Suite 300, Boca Raton, FL 33487-2742

and by CRC Press
4 Park Square, Milton Park, Abingdon, Oxon, OX14 4RN

CRC Press is an imprint of Taylor & Francis Group, LLC

Library of Congress Cataloging-in-Publication Data
Names: Gosser, David K., author.
Title: R for quantitative chemistry / David Gosser.
Description: First edition. | Boca Raton : CRC Press, 2024. |
Includes bibliographical references.
Identifiers: LCCN 2023005880 (print) | LCCN 2023005881 (ebook) |
ISBN 9781032414799 (hardback) | ISBN 9781032415475 (paperback) |
ISBN 9781003358640 (ebook)
Subjects: LCSH: Analytical chemistry—Quantitative—Data processing. |
R (Computer program language)
Classification: LCC QD101.2 .G68 2024 (print) | LCC QD101.2 (ebook) |
DDC 543.0285/5133—dc23/eng/20230531
LC record available at https://lccn.loc.gov/2023005880
LC ebook record available at https://lccn.loc.gov/2023005881

ISBN: 9781032414799 (hbk)
ISBN: 9781032415475 (pbk)
ISBN: 9781003358640 (ebk)

DOI: 10.1201/9781003358640

Typeset in Times
by codeMantra

Contents

Preface

This book originated from an effort to integrate programming into the teaching of Quantitative Chemistry (or Quantitative Chemical Analysis) at the City College of New York. In choosing a language, I found in R the following desirable qualities for teaching:

1. A comprehensive online platform in R Studio Cloud (now rebranded as Posit)
2. Easy to get up & running.
3. Quality graphics
4. Wide range of statistical packages.
5. Commonly used in scientific research
6. Vibrant online community

Each chapter introduces one or more problems that can be solved in R and includes a suggested project (typically involving student-collected or literature data) through which the student can deepen their understanding of analytical chemistry and practice new R skills.

I would like to acknowledge the helpful feedback from my students in Chemistry 243 at City College, and especially Aleksandr Knyazev, who formatted this book in latex and contributed the automated R program to perform Monte Carlo analysis with the subplex package.

Author

My original training in programming traces back to FORTRAN and punchcards & Mainframe at Brown University, where I completed a PhD in Chemistry & developed a simulation program to analyze electrochemical experiments, soon ported to PC with the lovely Turbo Pascal (now still available using FreePascal).

Intro to R

1

Every programming language has similar elements. We need to input data, identify variables, perform calculations, create functions, control program flow, and output data and graphics. The R language is attractive in that it has a full online development environment -**RStudio Cloud** (https://rstudio.cloud)- and built-in functions for regression analysis, solving equations, and producing graphics. R is increasingly used for those involved in data analysis and statistics, particularly social sciences, bio-statistics, and medicine. References on R tend to emphasize different features of R more suitable for social sciences, data mining, ecology, and medicine. A good reference for scientists using numerical methods is available [1], and an Analytical Chemistry textbook [2] that provides examples in R is available. This manual will present the elements of R gradually, as needed, mostly through chemical data examples. This document itself is created with **R Markdown** (https://bookdown.org/yihui/rmarkdown/), which integrates a document creation with R code.

1.1 R IN RSTUDIO CLOUD

We will write our programs on an online platform (an IDE) called Rstudio Cloud (RStudio Cloud is being renamed Posit Cloud), which makes it platform-independent - we access the program the same way on a PC, Mac, or Chromebook. The Rstudio Cloud environment is divided into four main sections, with the most important elements listed below.

- **Top Left**: Script - where you write R code
- **Bottom Left**: Console - show output
- **Top Right**: Environment - we will mostly ignore this
- **Bottom Right**: show Plots, help, packages

DOI: 10.1201/9781003358640-1

To start writing a program in Rstudio Cloud: create a **Space** (like a file folder that can contain related programs), select **New project**, and from the menu:
    File —— Newfile —— Rscript

& start typing! To run a program, highlight the code and select "run". You can create multiple **Projects** and R scripts in a single Space. Spaces are displayed on the left and you can create multiple spaces. In the tools icon (upper right), spaces can be shared. When you run a script program, results and error messages will appear on the console, and plots appear on the plot area. In this document, the shaded area is R code, and the output appears in a white background following the shaded R script. The theory is described in the text, and the details of R code are described in the code comment (text preceded by #)

1.2 VECTORS AND NUMERICS

In R we use $< -$ to set a variable to a value. R uses "vector" calculation, which avoids use of loops in more traditional programming. For instance, below we set x equal to a series of values (a vector), and then calculate a series of values of y.

```
#  A comment line


x <- 5.0
#  Set x equal to 5.0


y <- x^2
# y is equal to x squared.


x <- c(1.0, 2.0, 3.0)
# x is equal to a numbered list of values - a "vector"


x <- seq(1,2,0.2)
# create an incremented sequence


xx <-matrix(c(2,4,6,8),2,2)
# a matrix row 1 = 2 6 row 2 = 4 8
#  length(x) returns the length of the vector x.
#  x returns all the values of x


length(x)
```

```
x
```

```
## [1] 1.0 1.2 1.4 1.6 1.8 2.0
```

```
#  length(x) returns the length of the vector x.
#  x returns all the values of x
xx
```

```
##       [,1] [,2]
## [1,]    2    6
## [2,]    4    8
```

```
#  exponents and logs
```

```
10^(-2)
```

```
## [1] 0.01
```

```
# power of 10

exp(1)
```

```
## [1] 2.718282
```

```
# power of 2.718282 .....

log(10)
```

```
## [1] 2.302585
```

```
# base e

log10(10)    # base 10
```

```
## [1] 1
```

Once we have a series of *x* and *y*, of equal length, we can easily create a graph.

```
x    <-   seq(0,10,0.5)
# a sequence from 1 to 10, increments 0f 0.5


y <-   x^2
#  Note that y is calculated for every x.
#  This is called vectorized.


plot(x,y)
```

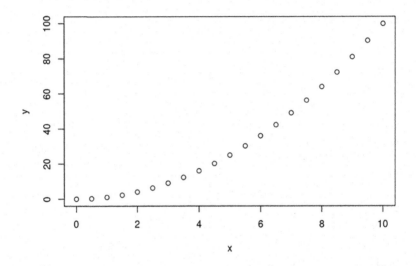

```
# Create a plot. We can add a lot of formatting!
```

Below is an example of basic formatting commands.

```
plot(x,y,type = "b",main = "A Formatted Graph",
col = "darkblue", xlab = "X Label", ylab = "Y Label")
```

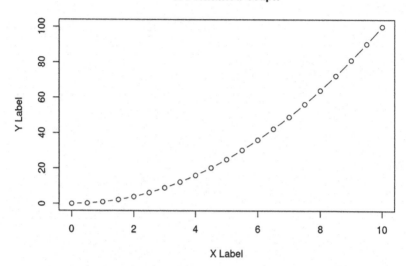

type="p": for points (by default)
type="l": for lines
type="b": for both; points are connected by a line

1.3 FUNCTIONS

We can define a function and evaluate it later in the R script.

```
Afunc <- function(x)  {x^2 + log(x)}


y <- c(1,2,3)


Afunc(y)



## [1]  1.000000  4.693147 10.098612



# note the vectorized evaluation
```

1.4 READING AND WRITING DATA FILES

With larger data sets, as in a titration, we prefer to be able to read a data file, such as output from a spreadsheet. A common filetype is csv (comma-separated values) files. Usually, the first row contains the headers, and the remaining rows the values. If you upload a csv file into your working directory, it can be read into dataframes. Dataframes are columns of values that do not have to be of the same type (Names, dates, income, etc), and they are very important in many R applications. However, here' we are focusing on vectors, and we can convert a dataframe with two or more numeric columns (such as volume and pH) into respective vectors. The data file "mydat.csv" looks like this:

vol, pH
1.5, 3.2
3.0, 6.6
7.0, 14.0

```
Mydata <- read.csv("mydat.csv")
# a file with headers vol and pH


Mydata
```

```
##    vol   pH
## 1 1.5   3.2
## 2 3.0   6.6
## 3 7.0  14.0
```

```
#  it's a dataframe


volume <- Mydata$vol
# extract column vol as vector


pH <- Mydata$pH
# extract column pH as vector


volume
```

```
## [1] 1.5 3.0 7.0
```

```
pH
```

```
## [1]  3.2  6.6 14.0
```

```
#  and to write a vector file
```

```
write(volume,file="newfile")
```

1.5 CONTROL STRUCTURES: FOR LOOP

Control structures allow iterative operations (for loop) and decision directed operations (if then else).

```
x <- 2

for(i in 1:4)   { x <- x*x}

x

## [1] 65536
```

1.6 PACKAGES

So far we have been using the base R that is available by creating an R script. However, we will also make use of R packages which extend R by adding graphical or computational extensions to R. They can be installed by selecting "Packages" and "Install" in the lower right quadrant of RStudio and searching by name for the package. Examples of packages we will use are nls2 (nonlinear regression), fftw (Fourier transform), and Stats (Statistics). The use of each package will be explored in the context of its use.

1.7 QUESTIONS AND PROJECTS

1. Create a vector "myvec" equal to $1, 5, 10, \ldots, 100$ with the sequence command, and create a function yval equal to myvec squared. Create a plot of yval versus myvec, with the data symbols as points.
2. Write the vector yval to a file. Check that the file was created.
3. $pH = -\log[H^+]$. Create a vector "pH" of pH values from 1 to 10. Create a function "hplus" which converts pH to hydrogen ion concentration. Call the function to calculate the hydrogen ion concentration for the series of pH values.

4. Find an equation with dependent and independent variables used in science or social science $(y = f(x))$. Create a reasonable range and sequence of x values, calculate the y values. Make a plot accurately labeled and with a title.

REFERENCES

1. O. Jones, R. Maillardet, and A. Robinson (2009). *Introduction to Scientific Programming and Simulation Using R*. CRC Press: Boca Raton, FL.
2. D. Harvey. Analytical Chemistry. https://chem.libretexts.org/Bookshelves/ Analytical_Chemistry/Analytical_Chemistry_2.1_(Harvey).

Data and Statistics

2

Every measurement has **random error**: Error that is inherent in the nature of the measurement. For instance, while the true mass of an object might be 1.0000 g, the actual measured mass will deviate from that value in a random manner, and the measured mass could well be 1.02 or 0.98 g. These deviations follow a Gaussian distribution. Random error is also referred to as noise. We can simulate the results of a number of measurements with the use of the R command rnorm. Below we simulate five measured values assuming a true value of 1. The rnorm command is useful in simulating realistic data. Using set.seed will reproduce the same set of random numbers.

DOI: 10.1201/9781003358640-2

```
noisy1 <- rnorm(5,0,0.2)

noisy1
```

```
## [1]  0.005311459  0.069111160  0.299931297 -0.261167773 -0.210372187
```

```
noisey2 <- 1 + rnorm(5,0,0.2)

noisey2
```

```
## [1] 0.6051724 1.0967651 0.9541377 0.9828683 1.1702550
```

```
set.seed(1)
# if set.seed(1) then gives same sequence

noisey3 <- 1.0 + rnorm(5,0,0.2)

noisey3
```

```
## [1] 0.8747092 1.0367287 0.8328743 1.3190562 1.0659016
```

rnorm(n,mean=0,sd=1) Generates Gaussian Random Values
input n is the number of values, mean, standard deviation. Default values are 0 for mean and standard deviation of 1.

Comments Most used here to simulate realistic data by adding noise to an idealized model. If we want a reproducible simulation we can use set.seed.

Significant Figures: The number of significant figures is all the certain figures plus one uncertain figure. The degree of uncertainty in the last digit is ultimately determined by a statistical analysis.

The **Mean** (or average) of a set of n measurements x is defined:

$$\bar{x} = \frac{\sum_{i=1}^{n} x_i}{n}$$

The mean and standard deviation can be related to the **Gaussian distribution** - that gives the probability of observing a particular value of x. For a finite number of measurements, the Gaussian distribution can be approximated as:

$$y = \frac{1}{s\sqrt{(2\pi)}} e^{\frac{-(x-\bar{x})^2}{2s^2}}$$

Data that we can collect from individuals or society can roughly follow a Gaussian distribution. In a study [1], the average male weight is 76.7 kg with a standard deviation of 12.1 kg, while women had an average weight of 61.5 with a standard deviation of 11.1 kg. Substituting this into the Gaussian distribution formula

```
s <- 12.1
# standard deviation


meanw <-  76.7
# mean weight

xval <- seq(20,100,2)

xval
```

```
## [1]  20  22  24  26  28  30  32  34  36  38  40  42  44  46  48  50  52  54  56
## [20] 58  60  62  64  66  68  70  72  74  76  78  80  82  84  86  88  90  92  94
## [39] 96  98 100
```

```
# here we calculate a gaussian distribution


yp <- (1/(s*sqrt(2*pi)))*exp(1)^((-(xval-meanw)^2)/(2*s^2))


yp
```

```
##  [1] 5.623116e-07 1.203433e-06 2.506117e-06 5.078268e-06 1.001301e-05
##  [6] 1.921095e-05 3.586475e-05 6.515106e-05 1.151622e-04 1.980767e-04
## [11] 3.315062e-04 5.398645e-04 8.554856e-04 1.319093e-03 1.979124e-03
## [16] 2.889386e-03 4.104619e-03 5.673812e-03 7.631536e-03 9.988120e-03
## [21] 1.272010e-02 1.576275e-02 1.900677e-02 2.230075e-02 2.546041e-02
## [26] 2.828436e-02 3.057469e-02 3.215974e-02 3.291531e-02 3.278070e-02
## [31] 3.176678e-02 2.995457e-02 2.748450e-02 2.453847e-02 2.131778e-02
## [36] 1.802068e-02 1.482297e-02 1.186409e-02 9.239917e-03 7.002235e-03
## [41] 5.163452e-03
```

```
plot(xval,yp,xlab="Weight in kg",ylab="Probability")
```

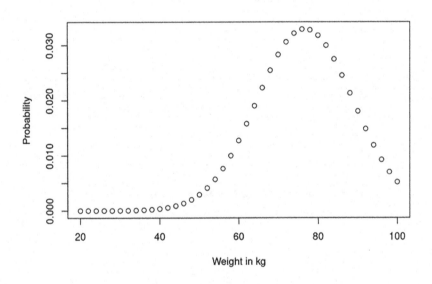

```
# R has a command for Gaussian distribution, dnorm.

dp  <- dnorm(xval,meanw,s)

plot(xval,dp,xlab="Weight in kg",ylab="Probability",ylim=c(0,.04),
     main="Women (x) and Men (o) Weight Distributions")

fdp <-   dnorm(xval,61.5,11.1)
# women weight distribution

points(xval,fdp,pch=4)
```

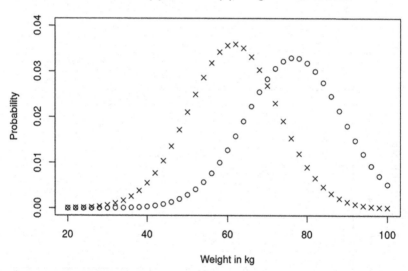

Note: We can specify symbols in the plot command. pch = 0,square
pch = 1,circle
pch = 2,triangle point up
pch = 3,plus
pch = 4,cross
pch = 5,diamond

Each point represents the probability of a particular observation, and the area under the curve (the sum of all probabilities) is 1. The distance from the mean of a particular measurement can be discussed in terms of "deviations from the mean" as multiples of the standard deviation. For instance: 68.3% of measurements lie within plus or minus one standard deviation 95.5% within plus or minus two standard deviations 99.7% within plus or minus three standard deviations. The R command for the cumulative distribution, which approaches 1, is pnorm.

```
cdp <-  pnorm(xval,76.7,12.1)
#   men  weight distribution
fdp <-  pnorm(xval,61.5,11.1)
#   women weight distribution

plot(xval,cdp,xlab="Weight",ylab="Cumulative Probability")
points(xval,fdp,pch=4)
```

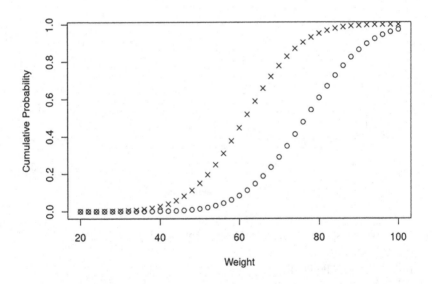

pnorm(n,mean=0,sd=1) Generates Gaussian Random Values
input: n is the number of values, mean, standard deviation. Default values are

0 for mean and standard deviation of 1.

Comments: Most used here to simulate realistic data by adding noise to and idealized model.

2.1 COMPARING TWO SETS OF DATA: THE t TEST

Argon was discovered because the mass of chemically generated nitrogen was significantly different from nitrogen obtained from air. Significant, in statistics, is carefully defined in terms of probabilities. An interesting example comes from history. Rayleigh's investigation [2] of the mass of chemically generated nitrogen (for instance, from the decomposition of pure NO) and nitrogen obtained from air was about 0.5% greater than that obtained from chemical decomposition. Was this slight difference attributable to experimental error? The mass of the two sets of masses were as follows, and the results of a t-test in R are shown.

```
Rdata <- c(2.31017,2.30986,2.31010,2.31001,2.310024,2.31010,2.31028)
Tdata <- c(2.30143,2.29890,2.29816,2.30182,2.29869,2.29940,2.29848)

Xdata <- data.frame(Rdata,Tdata)
knitr::kable(Xdata[,], col.names = c('Chem Data',
'Air Data'), caption = "Gas Data for t.test/grams")
```

Table 2.1: Gas Data for t.test/grams

Chem Data	Air Data
2.310170	2.30143
2.309860	2.29890
2.310100	2.29816
2.310010	2.30182
2.310024	2.29869
2.310100	2.29940
2.310280	2.29848

```
t.test(Rdata,Tdata)
```

```
##
## Welch Two Sample t-test
##
## data:  Rdata and Tdata
## t = 18.876, df = 6.0976, p-value = 1.218e-06
## alternative hypothesis: true difference in means is not equal to 0
## 95 percent confidence interval:
##  0.009164539 0.011882319
## sample estimates:
## mean of x mean of y
##  2.310078  2.299554
```

The important value to look for is p (probability) value: it tells us the probability that this overlap is due to merely random experimental error. The p value reported indicates about a one in a million chance that these results would occur if the masses were actually the same. We can also see that the size standard deviations, and see that they are much smaller than the difference between the two average values.

2.2 QUESTIONS AND PROJECTS

1. Simulate ten random values centered around 1, with a standard deviation of 0.1, 5 times. Compare the mean values obtained in each case. In each case, what is the standard deviation? What is the average standard deviation?

2. Gaussian Project
   - Find two data sets that are likely to follow random distribution and are likely to have similar mean values. This could be baseball batting averages, heights, or some other attribute. Should be >15 values.
   - Using R, create two vectors that represent the two data sets.
   - Find the mean and the standard deviation. The commands are - if X is the vector: mean(X) and sd(X).
   - Use the dnrom command and graph the estimated Gaussian distribution over a reasonable range.
   - Use the t-test to determine if they are significantly different. Interpret the results.

REFERENCES

1. W. Millar (1986). Distribution of body weight and height. *Journal of Epidemiology and Community Health*, 40: 319–323.
2. L. Rayleigh and W. Ramsay (1895). Argon, a new constituent of the atmosphere. *Philosophical Transactions of the Royal Society of London A*, 186: 187–241.

Beer's Law and Linear Regression

3

Beer's law for UV-Vis Spectroscopy is:

$$Abs = \epsilon_\lambda bc$$

where A = absorbance ϵ_λ = absorption coefficient at a particular wavelength, b = pathlength (typically cm), and c = concentration (usually in molarity or μ grams)

There is a linear relationship between the absorbance and the concentration, given a fixed pathlength and wavelength. In any experimental situation, the inherent noise in the data will mean that we need to include a constant (I = intercept).

$$Abs = \epsilon_\lambda bc + I$$

In order to use Beer's Law to analyze for a particular component in solution (an ãnalyte), we construct a calibration curve using experimentally determined data. The object is to determine the absorption coefficient, ϵ. We prepare solutions of varying concentrations of the analyte, and measure the absorbance. According to Beer's Law, a plot of Absorbance versus concentration should result in a straight line with a slope equal to ϵ.

Let's take an example. The data comes from the widely used Bradford assay for protein [1]. In the Bradford assay, a dye (Coomassie brilliant blue G-250) changes from red to blue when attached to proteins, and this color absorption at 595 nm) is used to measure protein concentration. Here is sample data for a calibration [2] for the BSA (Bovine Serum Albumin) standard.

DOI: 10.1201/9781003358640-3

```
Conc <- c(2.0, 5.0, 10.0, 15.0, 18.0)

# in micrograms/ml

Abs <- c(0.115,0.266,0.413,0.701,0.811)

# Absorbance


Xdata <- data.frame(Conc,Abs)

# x and y to a data frame


knitr::kable(Xdata[,],

col.names = c('Conc(microg/ml)','Abs'),

caption = "Bradford assay data")    # makes a table
```

Table 3.1: Bradford assay data

Conc(microg/ml)	Abs
2	0.115
5	0.266
10	0.413
15	0.701
18	0.811

3.1 LINEAR REGRESSION

We use the "lm" command to do linear regression - to find the best fit to the data according to the *least squares* criterion: the slope (ϵ) and intercept are adjusted to minimize the sum of the squares of the deviations of the experimental data from the fitted line. The concept of regression is used throughout this book and is central to analyzing data in chemistry. Mathematical models in Chemistry have adjustable parameters - for Beer's Law they are the absorptivity coefficient (the slope) and the intercept. Regression methods find the "best fit" line: the choice of slope and intercept which generates a line which will have the smallest sum of the squares of the deviations from the experimental data.

```
BL <- lm(Abs ~ Conc)
# Linear regression
```

```
summary(BL)
```

```
##
## Call:
## lm(formula = Abs ~ Conc)
##
## Residuals:
##      1      2       3      4      5
##  0.0018  0.0223 -0.0482  0.0223  0.0018
##
## Coefficients:
##              Estimate Std. Error t value Pr(>|t|)
## (Intercept) 0.026200   0.029054   0.902 0.433649
## Conc        0.043500   0.002495  17.435 0.000411 ***
## ---
## Signif. codes:  0 '***' 0.001 '**' 0.01 '*' 0.05 '.' 0.1 ' ' 1
##
## Residual standard error: 0.03329 on 3 degrees of freedom
## Multiple R-squared:  0.9902, Adjusted R-squared:  0.987
## F-statistic:   304 on 1 and 3 DF,  p-value: 0.0004113
```

```
# summarize results of linear regression

 cf <- coef(BL)
int <- cf[1]
sl <-  cf[2]
int
```

```
## (Intercept)
##      0.0262
```

```
sl
```

```
##    Conc
## 0.0435
```

```
# vector: intercept and slope

plot(Conc, Abs, main = "Beer's Law ", xlab = "",
ylab = "Absorbance", xlim = c(0.0,20.0),
ylim =   c(0.0, 1.0),
sub = "Figure 1. Beer's Law Plot for BSA")
title(xlab="Concentration, ug/ml", line=2, cex.lab=1.0)
library(plotrix)
abline(BL,col = "red")
ablineclip(a =0.0, b = 0.03,x1=1,x2=18)
```

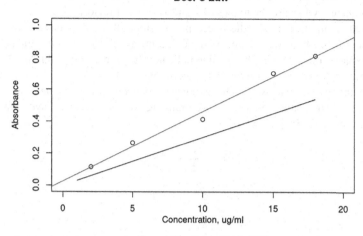

Beer's Law

Figure 1. Beer's Law Plot for BSA

```
ablineclip
```

```
# adding lines that show other slopes
```

Coomassie Blue. Associates with Proteins through non-covalent interactions.

From summary(BL) we learn the best fit values for the slope (ϵ) and the intercept, the standard deviations of these fitted values, the p values, and the r^2 value: how well a linear model explains the data.

lm(ydata,xdata) Linear regression on a set of paied x and y values.
input xdata and ydata are the names of x and y vectors previosly defined.
Comments Output of lm are the intercept and slope of the best fit line, the standard error of these values, and r squared, a measure of how well a linear model explains the data, and the p value indicates the probability pf thr "null" hypothesis: there is no linear relationship between x and y.

The final step in an analysis is to calculate the concentration of an unknown based on the absorbance measured. Rearranging Beer's Law, we have, for example, for an absorbance = 0.350

$$c = \frac{Abs - I}{\epsilon}$$

```
val <- (0.350 - cf[1] )/cf[2]
val
```

```
## (Intercept)
##      7.443678
```

```
# the unknown concentratiom
```

3.2 RESIDUAL PLOTS

In a regression analysis, it can be helpful to plot the residuals (the deviations) of the experimental data from the theoretical best fit line (or curve). The residuals should appear randomly distributed above and below the best fit line.

For the Bradford determination, we have:

```
res <- residuals(BL)
plot(Conc,res,ylab="residuals",ylim=c(0.05,-.05),
    main="Residual Plot Bradford Method")
abline(a = 0.0, b = 0.0, col = "blue")
```

Residual Plot Bradford Method

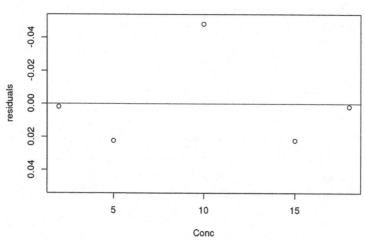

We can simulate an example with more points, based on a typical Bradford assay. The simulated data exaggerates the magnitude of typical deviations to help visualize.

```
Scon <- seq(0.5,20,0.5)
length(Scon)
```

```
## [1] 40
```

```
rnoise <- rnorm(40,0,0.1)
length(rnoise)
```

```
## [1] 40
```

```
Sabs <- 0.0435 * Scon + 0.026 + rnoise

plot(Scon,Sabs)

SBL <- lm(Sabs ~ Scon)

summary(SBL)
```

```
##
## Call:
## lm(formula = Sabs ~ Scon)
##
## Residuals:
##       Min        1Q    Median        3Q       Max
## -0.192232 -0.056654  0.002457  0.070929  0.207711
##
## Coefficients:
##             Estimate Std. Error t value Pr(>|t|)
## (Intercept) 0.036075   0.029980   1.203    0.236
## Scon        0.044481   0.002549  17.453   <2e-16 ***
## ---
## Signif. codes:  0 '***' 0.001 '**' 0.01 '*' 0.05 '.' 0.1 ' ' 1
##
## Residual standard error: 0.09303 on 38 degrees of freedom
## Multiple R-squared:  0.8891, Adjusted R-squared:  0.8862
## F-statistic: 304.6 on 1 and 38 DF,  p-value: < 2.2e-16
```

```
abline(a = 0.052, b = 0.042, col = "blue")
```

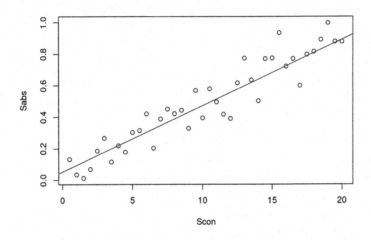

```
plot(Scon,residuals(SBL))

abline(a = 0.0, b = 0.0, col = "blue")
```

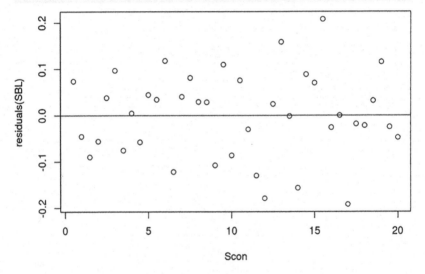

3.3 METHOD VALIDATION

There is much attention on method validation in the Analytical Literature and industries concerned with analyte measurement [3]. A brief summary is provided below:

- **Selectivity**: ability to measure in the presence of interferences.
- **Accuracy**: agreement with the true value.
- **Precision**: Given by relative standard deviation: the (standard deviation divided by the mean) times 100.
- **Linearity and Range**: over what concentrations does Beer's Law hold?
- **Limit of Detection**: At what concentration is the signal to noise ratio 3:1

3.4 QUESTIONS AND PROJECTS

Linear Regression Project

1. Find real world data that you expect to follow a linear relationship. Data points should be >10.
2. Enter the *x y* data as the vector "xval" and "yval" in R.
3. Use lm command to perform linear regression.
4. Summarize the results.
5. Plot the data and the best fit line.
6. Analysis & Interpretation: *p* value for intercept and slope.
7. Make a residual plot. Is it consistent with linear relationship?
8. What is the relative precision?

REFERENCES

1. M.M. Bradford (1976). Rapid and sensitive method for the quantitation of microgram quantities of protein utilizing the principle of protein-dye binding. *Analytical Biochemistry*, 72 (1–2): 248–254. https://www.sciencedirect.com/ science/article/abs/pii/0003269776905273?

2. P.N. Brady and M.A. Macnaughtan (2015). Evaluation of colorimetric assays for analyzing reductively methylated proteins: Biases and mechanistic insights. *Analytical Biochemistry*, 491: 43–51. https://www.ncbi.nlm.nih.gov/ pmc/articles/PMC4631703/.

3. T. Roa (1986). Validation of analytical methods. In: M.T. Stauffer (ed.), *Calibration and Validation of Analytical Methods*, vol. 40, pp. 319–323. BoD – Books on Demand. https://www.intechopen.com/chapters/57909.

Solving Equilibrium

4

4.1 SYSTEMATIC TREATMENT OF EQUILIBRIUM

In solving chemical equilibrium, we often make approximations to simplify the mathematical solution. For instance, we often use the approximation of negligible dissociation of weak acids. Here, we will explore the use of R to solve equilibrium without approximations. We find the equation to solve through application of a **systematic analysis of equilibrium**, which takes into account the equilibrium, charge balance, and mass balance expressions.

4.2 pH OF HYDROFLUORIC ACID

We can show four equations:

HF dissociation

$$HF \rightleftharpoons H^+ + F^-$$

$$K_a = \frac{[H^+][F^-]}{[HF]} = 6.8 \times 10^{-4}$$

Water Dissociation

$$H_2O \rightleftharpoons H^+ + OH^-$$
$$K_w = [H^+][OH^-] = 10^{-14}$$

Mass Balance

$$[HF]_i = [HF] + [F^-]$$

DOI: 10.1201/9781003358640-4

Charge Balance

$$[H^+] = [F^-] + [OH^-]$$

Successively substituting the mass balance, charge balance, and K_w into the dissociation expression: eliminate [HF]

$$K_a = \frac{[H^+][F^-]}{[HF]_i - [F^-]}$$

eliminate [F$^-$]

$$K_a = \frac{[H^+]([H^+] - [OH^-])}{[HF]_i - [H^+] + [OH^-]}$$

eliminate [OH$^-$]

$$K_a = \frac{[H^+]([H^+] - \frac{K_w}{[H^+]})}{[HF]_i - [H^+] + \frac{K_w}{[H^+]}}$$

Finally, simplifying a bit:

$$K_a \times ([HF]_i [H^+] - [H^+]^2 + K_w) - ([H^+]^3 - K_w [H^+]) = 0$$

We can solve this for [H$^+$] using the **uniroot** command.

```
Ka <- 6.8 * 10^-4

Kw <-  10^(-14)

CHF <- 0.0100  # Concentration of HF (Molarity)

Ka
```

```
## [1] 0.00068
```

```
Kw
```

```
## [1] 1e-14
```

```
HFFunc <- function(H) {

  Ka * (CHF*H - H^2 + Kw) -(H^3-Kw*H)

}

  Hroot <- uniroot(HFFunc,c(10^-6,10^-2),tol = 10^-18)

  Hroot
```

```
## $root
## [1] 0.002289753
##
## $f.root
## [1] 0
##
## $iter
## [1] 12
##
## $init.it
## [1] NA
##
## $estim.prec
## [1] 9.729952e-07
```

```
appx <- sqrt(Ka*0.01)   # the approximate solution
```

```
appx
```

```
## [1] 0.002607681
```

The solution is [H$^+$] 0.00229 M, or 23% dissiciated. Compare with the approximate solution, assuming negligible dissociation

$$[H^+] = \sqrt{K_a[HA]_i} = 0.00260M$$

uniroot(function,c(lower,upper),tol): Finds the root (zero) of a function given a lower and upper bound and a tolerance.
input: a function, upper and lower x values to bracket the search, and a tolerance (uncertainty in the final result)
Comments: important output is the root and the precision of the root.

4.3 SOLVING OVER A SERIES OF K_a VALUES WITH A FOR LOOP

We can use a for loop to iterate uniroot with a sequence of K_a values.

```
#   test a loop on uniroot


xroot <-  0   # declare xroot

                # initialize xroo
CHF <- 0.01
Kw <- 10^{-14}

HFFunc <-  function(H) {

  Ka * (CHF*H - H^2 + Kw) -(H^3-Kw*H)
}

for (i in 3:6)
{

    Ka <- 10^{-i}

    Hroot  <-  uniroot(HFFunc,c(10^-6,10^-2),tol = 10^-18)$root

    print(Hroot)

    xroot[i-2] <- Hroot

}
```

```
## [1] 0.002701562
## [1] 0.0009512492
## [1] 0.0003112673
## [1] 9.95013e-05
```

```
ka <- c(0.001,0.0001,0.00001,10^-6)
ka
```

```
## [1] 1e-03 1e-04 1e-05 1e-06
```

```
xroot
```

```
## [1] 0.0027015621 0.0009512492 0.0003112673 0.0000995013
```

```
plot(-log10(ka),xroot/CHF,xlab="pKa",
     ylab="fraction dissociation")
```

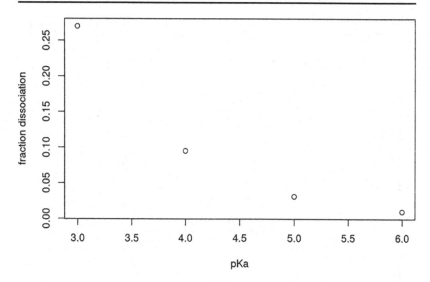

Of course, we could use the same method to integrate over a sequence of HF concentration values as well.

4.4 CARBON DIOXIDE AND ACIDIFICATION

Added carbon dioxide in atmosphere not only affects climate, but also pH of water. Let's explore using the systematic method. We assume aqueous carbon dioxide is fixed by the constant atmospheric carbon dioxide.

ATMOSPHERIC CARBON DIOXIDE (1960-2021)

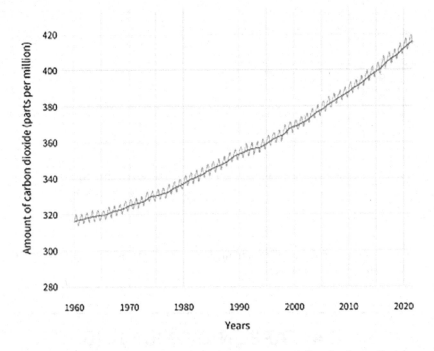

We start with Henry's law equilibrium between gas and dissolved gas. Carbon dioxide concentration is reported in ppm, so it is converted to atm.

$$CO_2(g) \rightleftharpoons CO_2(aq)$$

$$K_H = 0.034 = \frac{CO_2(aq)}{CO_2(g)}$$

$CO_2(g)$ is 415 ppm, or 0.000415 as a fraction
  We convert using:

$$Partial\,pressure = Total\,pressure \times mole\,fraction$$

If 1 atm is standard, then if CO_2 is 0.000415 parts of the atm, then the CO_2 pressure = 1 atm × 0.000415.

$$[CO_2](aq) = K_H[CO_2](g)$$

$$[CO_2](aq) = 0.034 \times 0.000415 = 1.41 \cdot 10^{-5} M$$

Given $CO_2(aq)$ is constant, we have four coupled equilibriums to consider:

$$CO_2(aq) + H_2O \rightleftharpoons H_2CO_3(aq)$$

$$K_{hyd} = 0.0026$$

hyd (hydrolysis)

$$H_2CO_3(aq) \rightleftharpoons HCO_3^-(aq) + H^+$$

$$K_{a_1} = 1.67 \cdot 10^{-4}$$

$$HCO_3^-(aq) \rightleftharpoons CO_3^-(aq) + H^+$$

$$K_{a_2} = 5.6 \cdot 10^{-11}$$

$$H_2O \rightleftharpoons H^+ + OH^-$$

$$K_w = 1.00 \cdot 10^{-14}$$

Charge Balance

$$[H^+] = [OH^-] + [HCO_3^-] + 2[CO_3^{2-}]$$

Substitute the K_w, K_{a_1}, and K_{a_2} expressions into the charge balance

$$[H^+] = \frac{K_w}{[H^+]} + \frac{K_{a_1} \cdot [H_2CO_3]}{[H^+]} + \frac{2K_{a_2} \cdot [HCO_3^-]}{[H^+]}$$

Substituting K_{hyd} and K_{a_1} (again)

$$[H^+] = \frac{K_w}{[H^+]} + \frac{K_{a_1} \cdot K_{hyd} \cdot [CO_2]_{aq}}{[H^+]} + \frac{2 \cdot K_{a_1} \cdot K_{a_2} \cdot [H_2CO_3]}{[H^+]^2}$$

Finally, K_{hyd} again,

$$[H^+] = \frac{K_w}{[H^+]} + \frac{K_{a_1} \cdot K_{hyd} \cdot [CO_2]_{aq}}{[H^+]} + \frac{2K_{a_1} \cdot K_{a_2} \cdot K_{hyd} \cdot [CO_2]_{aq}}{[H^+]^2}$$

Multiply by $[H^+]^2$ and total to

$$[H^+]^3 - K_w \cdot [H^+] + K_{a_1} \cdot K_{hyd} \cdot [CO_2]_{aq} \cdot [H^+] + 2K_{a_1} \cdot K_{a_2} \cdot K_{hyd} \cdot [CO_2]_{aq} = 0$$

An approximate approach:

Since K_{a_2} and K_w are small compare to K_{a_1}, we could try to ignore them, and we would have:

$$K_{a_1} = [H^+]^2 / H_2CO_3$$

where
$$H_2CO_3 = K_{hyd} \cdot CO_2(aq) = 0.0026 \cdot 1.41 \times 10^{-5}$$

And
$$[H^+]^2 = K_{a_1} \cdot K_{hyd} \cdot CO_2(aq)$$

```
Kh <- 0.0034
Khyd <-  0.0026
Ka1  <-  1.67 * 10^-4
Ka2  <-   5.6 * 10^-11
Kw   <-  1.00 * 10^-14
ca <-  1.41 * 10^-5                # concentration of co2(aq)


Kh
```

```
## [1] 0.0034
```

```
Khyd
```

```
## [1] 0.0026
```

```
Ka1
```

```
## [1] 0.000167
```

```
Ka2
```

```
## [1] 5.6e-11
```

```
Kw
```

```
## [1] 1e-14
```

```
ca
```

```
## [1] 1.41e-05
```

```
froot3 <- function(H)

{H^3 - Kw*H - Ka1 * Khyd  * ca * H - 2.0 * Ka1 * Ka2 * Khyd * ca}

X <- uniroot(froot3, c(10^(-6),10^(-4)), tol = 10^(-18))$root

X
```

```
## [1] 2.476388e-06
```

```
pH <-  -log10(X)

pH
```

```
## [1] 5.606181
```

4.5 VISUALIZING ROOT FINDING

The process of finding a root can be illustrated by evaluating the root function at pH values in the neighborhood of the 5.6. The root function can be seen to go from plus to minus in this region.

```
ph <- seq(5.5,5.7,0.02)

ph
```

```
## [1] 5.50 5.52 5.54 5.56 5.58 5.60 5.62 5.64 5.66 5.68 5.70
```

```
h <- 10^-ph

h
```

```
## [1] 3.162278e-06 3.019952e-06 2.884032e-06 2.754229e-06 2.630268e-06
## [6] 2.511886e-06 2.398833e-06 2.290868e-06 2.187762e-06 2.089296e-06
## [11] 1.995262e-06
```

```
val <- froot3(h)

val
```

```
## [1]  1.223031e-17  9.022593e-18  6.302128e-18  4.002739e-18  2.066941e-18
## [6]  4.448060e-19 -9.070142e-19 -2.026146e-18 -2.945236e-18 -3.692601e-18
## [11] -4.292791e-18
```

```
plot(ph,val)
```

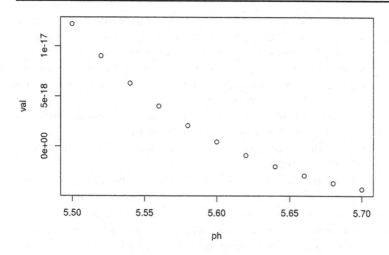

```
#    the pH of standing water is 5.6
```

The situation in natural waters is more complex and may involve further reactions of solid calcium carbonate. So ocean pH is not 5.6, but has values between 7 and 8. Nevertheless, atmospheric CO_2 is causing some acidification. calcium carbonate dissolution can mitigate ocean acidification, although is not an equilibrium process.

4.6 QUESTIONS AND PROJECTS

1. Consider a $HClO_2$ (chlorous acid0 with $K_a = 1.2 \cdot 10^{-2}$.
   Solve for H^+ using the systematic method, taking into account K_{b_1}, K_w, charge balance and mass balance. In solving this system, we can safely ignore K_{b_2}. Solve to find the pH of 0.01 M with uniroot and also use a for loop to explore the fraction of dissociation versus concentration of HA.

2. The Halfway Method for Solving Equilibrium Problems
   Based on the simplest root-finding method (bisection), here is a way of calculating a better answer when the approximation of slight dissociation is not satisfied. It is similar to the way a calculator or a computer solves equations (i.e. it is a numerical method). Another

method, cited in some Chemistry texts, is the method of successive approximations. Unfortunately, this method is unstable in some circumstances. In practice, there are faster methods, but they are more complicated. Additionally, this method requires understanding of the reaction quotient and its relationship to the equilibrium constant. The method can be summarized as follows:

- Define the extreme limits of dissociation. For instance, the 0.0100 M HA can be zero percent dissociated or 100
- Select the concentration value of H^+ "halfway" between 0.00 and 0.010 M, 0.0050 M
- Substitute that value of H^+ into the reaction quotient Q expression: $Q = \frac{[H^+]^2}{[HA]i-[H^+]} = (0.0050)20.0050 = 0.005$ note: Q is the same form as K_a but when concentrations are not at equilibrium.
- Compare the Q value to the K_a value. If $Q > K_a$ then the amount of dissociation is <0.0050 M. If $Q < K_a$, then the actual amount of dissociation is >0.0050 M. The concentrations must change so that Q becomes closer to K. In this case the $Q < 0.005$, and it therefore must be more dissociated than 0.005.
- Now the range must be somewhere between 0.005 and 0.010, and we choose a new "halfway" that is 0.0075. Our new Q is: $Q = \frac{(0.0075)^2}{0.0025}$ = 0.0225, and $Q > K$, and our next value must be a smaller dissociation. The next halfway is between 0.00500 and 0.0075, which is = 0.00625. The $Q = 0.010$, and we must go smaller again! The halfway is between 0.0050 and 0.00625, which is 0.005625, and $Q = 0.007$, which is pretty close to 0.0066. We can go as far as we want (or tell a computer to do it) and get as accurate as we want. Here we are satisfied to know that the dissociation has to be between 0.00500 and 0.00563, so we end with $H^+ = 0.0053$ M. By the way, what does the approximate method give: $H^+ = SQRT(0.0066 \times 0.01) = 0.00812$ M. This is much worse than the actual dissociation −80% versus 53%. So pH = 2.27

Now you try it with the case of aspartic acid, with a $pK_1 = 1.88$. What is the pH of 0.01 M aspartic acid?

Also: A good numerical method can "bracket" how close the answer must be to the true answer. The halfway method does this. Each step successively brackets the true value between smaller and smaller ranges.

Your R project: Translate the "halfway method" to an R program using a for loop to iterate and an if else statement to decide on the next calculation.

Titrations

5

Here we will extend the use of the R command "uniroot" to solve for an acid-base titration curve. First, we quickly review the common approximate method that is often presented.

5.1 THE APPROXIMATE VIEW

For the titration of a weak acid with a strong base, we can use approximate solutions for different parts of the titration curve. This is useful to get a qualitative understanding of the chemical equilibrium. For instance, the analysis of amino acids (diprotic or triprotic) is often considered in this light.

For now, let's consider the titration of 15.0 mL of 0.1 M lactic acid, $K_a = 1.38 \cdot 10^{-4}$ with 0.1 M NaOH.

5.1.1 Initial H

We use

$$[H^+] = \sqrt{K_a \cdot [HA]_i}$$

```
Ka <- 1.38 * 10^(-4)

cA <- 0.100

H <- sqrt(Ka * cA)

H
```

```
## [1] 0.003714835
```

```
pHi <- -log10(H)

pHi
```

```
## [1] 2.43006
```

5.1.2 The Buffer Region

In the buffer region (between 0 and equivalence point) we simply use the equilibrium expression and assume a complete acid-base reaction and no dissociation.

$$[H^+] = K_a \cdot \frac{[HA]}{[A^-]}$$

or

$$pH = pK_a + \log\frac{[A^-]}{[HA]}$$

We can keep this approximation in mind as we explore more accurate solutions.

```
pHmid <- -log10(1.38*10^{-4})
pHmid
```

```
## [1] 3.860121
```

which leads to the pH at the midpoint = pK_a

5.1.3 Endpoint

Here, we use the conjugate base and assume slight hydrolysis.

```
Kb <- 10^(-14)/(1.38*10^(-4))
Kb
```

```
## [1] 7.246377e-11
```

```
cBeq <- 0.1*0.015/0.035
cBeq
```

```
## [1] 0.04285714
```

```
oheq <- sqrt(Kb*cBeq)
oheq
```

```
## [1] 1.762268e-06
```

```
heq <- 10^(-14)/oheq
heq
```

```
## [1] 5.674504e-09
```

We can keep this approximation in mind as we explore more accurate solutions.

5.2 AN EXACT SOLUTION

Approaching in a systemic manner, we write down system equations for the weak acid/strong base titration, where VA= volume of acid, cA = Concentration of acid, cB = concentration of base, and the equilibrium constants K_a and K_w. In this way, we can get an expression of the volume of base as a function of [H^+] (or pH). This makes direct comparisons with experiment difficult, as we would normally want to predict pH as a function of added base.

Equilibrium Expressions

$$K_a = \frac{[H^+][A^-]}{[HA]}$$

$$K_w = [H^+][OH^-]$$

Charge Balance

$$[Na^+] + [H^+] = [A^-] + [OH^-]$$

Sodium Ion Concentration

$$\frac{VB \cdot cB}{VA + VB}$$

Mass Balance

$$[HA]_{tot} = [HA] + [A^-]$$

This is a bit tricky. We can easily substitute the sodium ion concentration and $[OH^-] = \frac{K_w}{[H^+]}$ to get:

$$\frac{VB \cdot cB}{VA + VB} + H^+ = A^- + \frac{K_w}{[H^+]}$$

Solving for [A^-] is a bit more work.

Substituting the mass balance into the equilibrium expression:

$$K_a = [A^-][H^+]/[HA] = \frac{[A^-][H^+]}{([HA]_{tot} - [A^-])}$$

$$K_a \cdot [\text{Ha}]_{\text{tot}} - K_a \cdot [\text{A}^-] = [\text{A}^-] \cdot [\text{H}^+]$$

$$K_a \cdot [\text{HA}]_{\text{tot}}/[\text{A}^-] = [\text{H}^+] + K_a$$

$$\frac{[\text{HA}]_{\text{tot}}}{[\text{A}^-]} = \frac{([\text{H}^+] + K_a)}{K_a}$$

let fA = fraction of A = $\frac{[\text{A}^-]}{[\text{HA}]_{\text{tot}}}$

$$fA = \frac{K_a}{K_a + [\text{H}^+]}$$

now the fraction of A- times the total conc HA is A- conc

$$[\text{A}^-] = fA \cdot \frac{cA \cdot VA}{Va + VB}$$

and we can substitute to get VB as function of other stuff

$$\frac{VB \cdot cB}{VA + VB} + H^+ = \frac{fA \cdot cA \cdot VA}{(VA + VB)} + \frac{K_w}{[\text{H}^+]}$$

this can be rearranged to:

$$VB = \frac{\left(K_a/([\text{H}^+] + K_a) - \left([\text{H}^+] - \frac{K_w}{[\text{H}^+]cA} \right) \right)}{\left(1.00 + \frac{H - K_w}{[\text{H}^+]cB} \right)} \cdot \frac{cA \cdot VA}{cB}$$

Below we use this expression to calculate a titration curve for a weak acid.

```
## volume as function pH

# Generating Titration Curve

#    pH f vol


cA = 0.100
# concentration acid Molar
cB = 0.100
#    base Molar
VA = 0.025   # volume of acid L
Ka = 1.38 * 10^(-4)
Kw = 1.00 * 10^(-14)


pH <- seq(3,12,0.5)
pH
```

```
## [1]  3.0  3.5  4.0  4.5  5.0  5.5  6.0  6.5  7.0  7.5  8.0  8.5  9.0  9.5 10.0
## [16] 10.5 11.0 11.5 12.0
```

```
#   a using sequence of pH values

length(pH)
```

```
## [1] 19
```

```
  H <- 10^-pH
# generate a sequence of H+ values
```

```
VB <- (Ka/(H + Ka) - (H - Kw/H)/cA) /
(1.00 + (H - Kw/H)/cB) * cA*VA/cB
```

```
VB
```

```
## [1] 0.002754094 0.007492558 0.014456342 0.020324917 0.023305981 0.024438395
## [7] 0.024819651 0.024942701 0.024981897 0.024994415 0.024998684 0.025001007
## [13] 0.025004819 0.025015759 0.025050032 0.025158610 0.025505049 0.026632771
## [19] 0.030555555
```

```
pH
```

```
## [1] 3.0 3.5 4.0 4.5 5.0 5.5 6.0 6.5 7.0 7.5 8.0 8.5 9.0 9.5 10.0
## [16] 10.5 11.0 11.5 12.0
```

```
plot(VB,pH,main = "Weak Acid Titration",
     xlab = "Volume Base / L", xlim = c(0,0.035),ylim = c(2,13) )
```

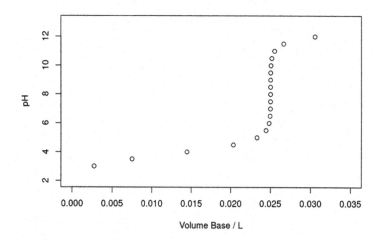

5.3 ROOT FINDING: pH AS FUNCTION OF VOLUME

```
cA = 0.100
# concentration acid Molar
cB = 0.100
# base Molar
VA = 25.00
# volume of acid ml
Ka = 1.00 * 10^-4
Kw = 1.00 * 10^-14


VB = 12.500  # midpoint test


froot <- function(H)  {


   VB - (Ka/(H + Ka) - (H - Kw/H)/cA) /
   (1.00 +  (H - Kw/H)/cB) * cA*VA/cB    }


 X <- uniroot(froot, c(10^(-12),10^(-3)), tol = 10^(-14))


X
```

```
## $root
## [1] 9.940534e-05
##
## $f.root
## [1] 3.215206e-12
##
## $iter
## [1] 10
##
## $init.it
## [1] NA
##
## $estim.prec
## [1] 5.000042e-15
```

5.4 USING SAPPLY WITH UNIROOT TO SOLVE FOR TITRATION CURVE

Since we have an implicit expression for the hydrogen ion conentration, we can use uniroot to solve for the hydrogen ion concentration, to generate a curve that can be more directly compared to the experiment. We could have used a for loop, but we can use the "sapply" command on uniroot.

```
# numerical soution of titration weak acid / strong base


#  used sapply to run unirooot for different values


#   generate full titration curve

cA = 0.100
# concentration acid Molar
cB = 0.100
#   base Molar
VA = 25.00
#  volume of acid ml
Ka = 1.38 * 10^(-4)
Kw = 1.00 * 10^(-14)


VB <- 0

 froot <- function(VB,H)

    { VB - (Ka/(H + Ka) - (H - Kw/H)/cA) /
          (1.00 +  (H - Kw/H)/cB) * cA*VA/cB  }

numb <- sapply(seq(0,30,1), function(VB) uniroot(froot, c(10^-12,10^-1),tol = 10^-16, VB=VB)$root)


 VB <- seq(0,30,1)
   numb
```

```
## [1] 3.646476e-03 2.095322e-03 1.325222e-03 9.209733e-04 6.837629e-04
## [6] 5.307068e-04 4.246551e-04 3.471577e-04 2.881837e-04 2.418625e-04
## [11] 2.045476e-04 1.738619e-04 1.481923e-04 1.264076e-04 1.076915e-04
## [16] 9.144053e-05 7.719929e-05 6.461771e-05 5.342237e-05 4.339665e-05
## [21] 3.436672e-05 2.619152e-05 1.875537e-05 1.196256e-05 5.733181e-06
## [26] 5.246435e-09 5.100009e-12 2.600000e-12 1.766671e-12 1.349997e-12
## [31] 1.099998e-12
```

```
length(numb)
```

```
## [1] 31
```

```
ph <- -log10(numb)
ph
```

```
## [1]  2.438127  2.678749  2.877711  3.035753  3.165094  3.275145  3.371964
## [8]  3.459473  3.540331  3.616431  3.689206  3.759796  3.829174  3.898227
## [15] 3.967819  4.038861  4.112387  4.189648  4.272277  4.362544  4.463862
## [22] 4.581839  4.726874  4.922176  5.241604  8.280136 11.292429 11.585027
## [29] 11.752844 11.869667 11.958608
```

VB

```
## [1]  0  1  2  3  4  5  6  7  8  9 10 11 12 13 14 15 16 17 18 19 20 21 22 23 24
## [26] 25 26 27 28 29 30
```

```
length(VB)
```

```
## [1] 31
```

```
length(ph)
```

```
## [1] 31
```

```
plot(VB,ph,xlab = "volume of base/ml", main = "Titration")
```

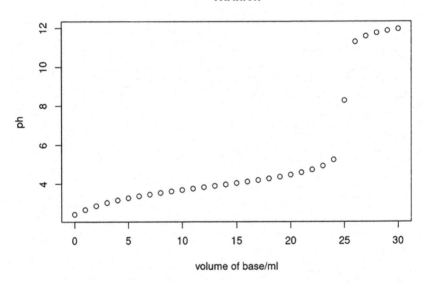

5.5 EFFECT OF IONIC STRENGTH/ACTIVITY COEFFICIENTS

Formally, we learn in physical chemistry that equilibrium constants are formulated with activities, not concentrations. In many instances, the distinction may not be important: for instance, we determine an equilibrium constant under specified conditions. In thermodynamics, equilibrium constants are defined as unit activity, which is obtainable at an ionic strength equal to zero. It does affect our experiments on binding curves. Kennedy [1] discusses a practical approach that we might apply to our analysis.

5.5.1 Ionic Strengths and Activity

Ionic Strength is defined:

$$I = \frac{1}{2} \sum c Z^2$$

where c is concentration and is charge.

The thermodynamic equilibrium constant for weak acid with charge z is

$$K_{th} = \frac{\alpha_{H^+}\alpha_{A^{z-1}}}{\alpha_{HA^z}}$$

the equilibrium measured at ionic strength = 0. (an ideality, which can be approached by doing experiments at a series of ionic strengths approaching 0 and extrapolating to zero)

where $\alpha_{H^+} = f[H^+]$

where f_{H^+} is the activity and alpha the activity coefficient

Because glass pH electrodes respond to proton activity, while HA and A- are measured in molarities, a mixed equilibrium constant is defined ref

$$K_{mix} = \frac{\alpha_{H^+}[A^{z-l}]}{[HA^Z]}$$

and

$$pK_{mix} = pK_{th} + \log f_{A^{z-1}} - \log f_{HA^z}$$

The Debye Huckel equation is useful up to $I = 0.005$

$$\log 10 f_i = -Az^2\sqrt{I}$$

under standard conditions $A = 0.509$

A more realistic equation is:

$$\log 10 f_i = \frac{-Az^2\sqrt{I}}{(1+\sqrt{I})}$$

and for the nth dissociation of acid with charge z

$$pK_{\text{mix},n} = pK_{\text{th},n} + \frac{A\sqrt{I}(2z - 2n + 1)}{1+\sqrt{I}}$$

We can use this equation to make an approximate correct K values obtained from titration experiments.

5.6 PROJECT

1. Using the pK value(s) determined from a titration, simulate the titration and compare (plot) both the titration curve and the simulation.
2. Apply activity theory to correct for ionic strength in determining K_a.

REFERENCE

1. C. Kennedy (1990). Ionic strength and the dissociation of acids, *Biochemical Education*, 18(1): 35–40.

Binding Curves

6

Binding Curves (or Binding Assays) are an approach to determining the equilibrium constant for the interaction of two molecules, commonly used in biochemistry [1]. Typically, a small molecule (commonly called the "ligand") attaches to a larger molecule such as a protein. If we denote the protein as P and the ligand as L we can express the dissociation as:

$$PL \rightleftharpoons P + L$$

and where the total protein is the bound plus unbound.

$$P_{tot} = PL + P$$

With the dissociation constant:

$$K_d = \frac{[P][L]}{[PL]}$$

The concentration of the ligand (in excess) is varied and the fraction bound $\frac{[PL]}{P_{tot}}$ is plotted versus the L [2].

Substituting $P_{tot} = P + PL$ into the K_d expression:
we get

$$K_d = \frac{([P_{tot}] - [PL])[L]}{[PL]} = \frac{([P_{tot}][L] - [PL][L])}{[PL]}$$

$$K_d[PL] + [PL][L] = [P_{tot}][L]$$

$$[PL](K_d + [L]) = [P_{tot}][L]$$

and we get expression for fraction bound as a function of K_d and L.

$$\frac{[PL]}{P_{tot}} = \frac{[L]}{K_d + [L]}$$

DOI: 10.1201/9781003358640-6

note as L increases the fraction bound goes from 0 to 1.

The key to binding assay is to find an experimental method to determine the concentration of PL, the bound form.

Although it is called a "binding assay" the equilibrium is discussed in terms of the dissociation, with K_d

6.1 MYOGLOBIN BINDING

```
# Nonlinear Regression / Binding Curve
# simulate noisy data for myoglobin
# KD = 0.26 [L] in kPa
# data generated with random noise added


library(nls2)
```

```
## Loading required package: proto
```

```
KD <- 0.26             # initializing variables
L <- 0.0
y <- 0.0


L <- seq(.1,1.5,0.1)   # set of 15 ligand concentrations


rnd <- rnorm(15,0,0.04)  #   set of 15 random values


y <- L/(KD+L) + rnd
# binding curve function with noise


plot(L,y, main = "Simulated binding data", xlab = "pO2 Kpa",
     ylab = "Fractional Binding")
```

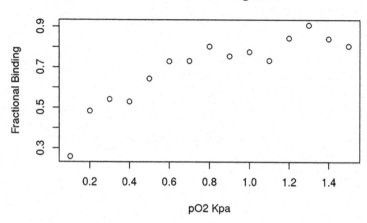

```
tryfit <- nls2(y ~ L/(KD+L),
                start = c(KD = 5))

summary(tryfit)
```

```
##
## Formula: y ~ L/(KD + L)
##
## Parameters:
##    Estimate Std. Error t value Pr(>|t|)
## KD 0.26347    0.01561   16.88 1.06e-10 ***
## ---
## Signif. codes:  0 '***' 0.001 '**' 0.01 '*' 0.05 '.' 0.1 ' ' 1
##
## Residual standard error: 0.04381 on 14 degrees of freedom
##
## Number of iterations to convergence: 7
## Achieved convergence tolerance: 7.89e-08
```

```
plot(L,y, main = "Myoglobin Binding Curve", xlab = "pO2 Kpa", ylab = "Fractional Binding")
```

```
lines(L,predict(tryfit), col = "blue")
```

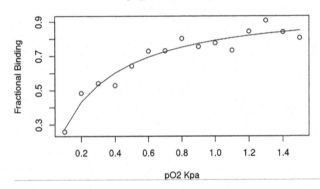

nls2$(y\ f(v, w, x)$, start $= c(v = 1, w = 2, x = 3)$): Non-linear regression with data vector y and function with parameters v, w, and x.

input: y data as vector, the function, and starting values for the parameters.

Comments: Output estimates of parameters with their standard errors. p value output may not be reliable nor nonlinear regression.

6.2 L NOT IN EXCESS

For the Binding Assay, with P = Initial Protein and L = Initial Ligand, and PL = x = amount of bound protein, we have the equilibrium expression:

$$K_d = \frac{([P] - x)([L] - x)}{x}$$

if we solve for x and divide by L, we will get fraction bound.

$$K_d = \frac{[P][L] - [P]x - x[L] + x^2}{x}$$

$$K_d x = [P][L] - [P]x - x[L] + x^2$$

$$[P][L] - [P]x - x[L] - K_d x + x^2 = 0$$

$$[P][L] - (K_d + [P] + [L])x + x^2 = 0$$

The quadratic solution, divided by [L],

$$\frac{x}{L} = \frac{(K_d + [P] + [L]) - \sqrt{(K_d + [P] + [L])^2 - 4([P] + [L])}}{2[L]}$$

note: why not divide by P? because P is large and L is limiting.

6.3 ACID/BASE BINDING CURVE

Adapt to titration data, framed as fraction binding as determined from the endpoint, vs pH (which is actually equilibrium ligand H)

We can transform titration data into a binding curve representation.

For the equilibrium:

$$HA \rightleftharpoons H^+ + A^-$$

$$K_a = \frac{[H^+][A^-]}{[HA]}$$

we are measuring H^+ and for a weak base strong acid titration, the fraction bound is fraction of the way to the endpoint.

$$f = \frac{[HA]}{[A^-] + [HA]} = \frac{vol}{vol_t}$$

substituting the equilibrium expressions into this:

$$f = \frac{[HA]}{\frac{K_a[HA]}{[H^+]} + [HA]}$$

$$\frac{1}{\frac{K_a}{[H^+]} + 1}$$

$$f = \frac{[H^+]}{[H^+] + K_a}$$

It is useful to note that we are measuring $[H^+]$, so unlike the similar binding equation, there is no approximation of excess "ligand".

This is a very useful way to analyze acid-base data to determine K_a values.

Where is the error? The error is in the estimation of the endpoint, which translates into a systematic error in the fraction bound.

```
KD <- 1.75 * 10^(-5)
L <- 0.0
y <- 0.0


pH <-  seq(3,6,0.2)  #   generating a list of values

length(pH)
```

```
## [1] 16
```

```
L <- 10^(-pH)

ra <-rnorm(16,0,0.005)

length(ra)
```

```
## [1] 16
```

```
y <-  L/(KD+L)

y <-  L/(KD+L) + ra

plot(pH,y)
```

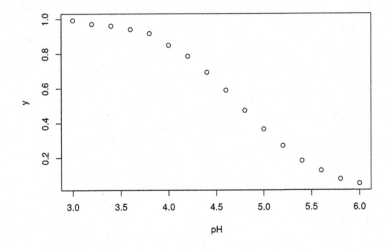

```
tryfit <- nls(y ~ L/(KD+L),
            start = c(KD = 0.0001))

summary(tryfit)
```

```
## Formula: y ~ L/(KD + L)
##
## Parameters:
##      Estimate Std. Error t value Pr(>|t|)
## KD 1.737e-05  1.620e-07   107.2   <2e-16 ***
## ---
## Signif. codes:  0 '***' 0.001 '**' 0.01 '*' 0.05 '.' 0.1 ' ' 1
##
## Residual standard error: 0.005593 on 15 degrees of freedom
##
## Number of iterations to convergence: 4
## Achieved convergence tolerance: 1.625e-07
```

```
plot(pH,y, main = "Acid/Base Binding", xlab = "pH",
     ylab = "Fractional Binding")

lines(pH,predict(tryfit), col = "blue")
```

Acid/Base Binding

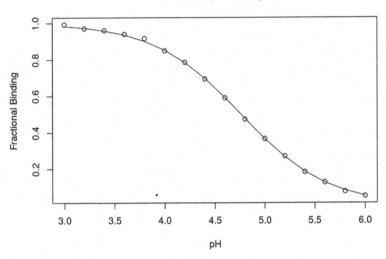

This approach to acid-base binding can be generalized, and the solution for a multi-protic system is:

$$f = \frac{\frac{[\text{H}^+]}{K_1} + \frac{2[H^+]^2}{K_1 K_2} + \frac{3[H^+]^3}{K_1 K_2 K_3}}{1 + \frac{[H^+]}{K_1} + \frac{[H^+]^2}{K_1 K_2} + \frac{[H^+]^3}{K_1 K_2 K_3}}$$

6.4 EXPLORATION OF DIPROTIC TITRATIONS

6.4.1 Preparation of Experimental Data

We can transform titration data to a binding curve format.
    if HA had slight dissociation we simply would have

$$F = (2 - \text{VA} \times \text{cB})/(\text{VE} \times \text{VB})$$

but in actuality, the fraction bound is less according to the extent of dissociation - the amount of H^+.

$$F <- (2 - ((VA \times cB) + ((H) \times (VI + VA)))) / (VE \times CB)$$

```
# example of diprotic acid fitting


pH <-read.csv("data3.csv")

Vol <- read.csv("data4.csv")


length(Vol)

## [1] 1


VolVect <- Vol$V2
# this extracts a vector from dataframe column

length(VolVect)

## [1] 100


pHVect <- pH$V1

length(pHVect)

## [1] 100


H <- 10^-(pHVect)
```

```
length(H)
```

```
## [1] 100
```

```
L <- H
```

```
plot(VolVect,pHVect,xlab = "Vol", ylab = "pH", main = "Titration")
```

Titration

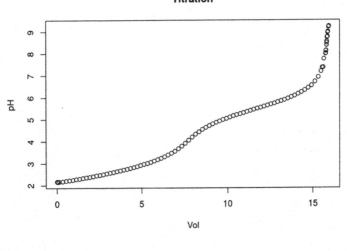

```
#Volume added at the 1st endpoint:
VE=7.905

#Initial volume of unknown acid:
VI=25

#The concentration of the base NaOH:
CB=0.10
CB
```

[1] 0.1

```
VA <- VolVect

F  <- (2-(((VA*CB)+((H)*(VI+VA)))/(VE*CB)))
#  fraction bound for each data point

tF <- F

length(pHVect)
```

[1] 100

```
length(F)
```

[1] 100

```
plot(pHVect,tF)
```

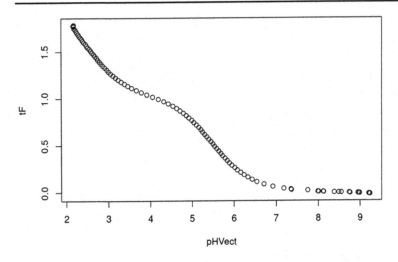

```
library(nls2)

# try with all the data

tryfit <- nls2(F ~ (L/KD1 + 2*L^2/(KD1*KD2))/(1+L/KD1 + L^2/(KD1*KD2) ),
                start = c(KD1 = 0.0001,KD2=0.01))

tryfit
```

```
## Nonlinear regression model
##   model: F ~ (L/KD1 + 2 * L^2/(KD1 * KD2))/(1 + L/KD1 + L^2/(KD1 * KD2))
##    data: <environment>
##       KD1        KD2
## 2.940e-06 2.426e-03
##  residual sum-of-squares: 0.02522
##
## Number of iterations to convergence: 7
## Achieved convergence tolerance: 3.722e-06
```

```
length((predict(tryfit)))
```

```
## [1] 100
```

```
plot(pHVect,F)
```

```
lines(pHVect,predict(tryfit), col = "blue")
```

6.5 PROJECTS

1. Transform monoprotic or diprotic data to a binding curve and analyze with nls2. Compare with the traditional midpoint analysis of a titration curve.

REFERENCES

1. T. Pollard (2017). A guide to simple and informative binding assays. *Molecular Biology of the Cell*, 21(23). doi: 10.1091/mbc.e10-08-0683.
2. While the mathematics of equilibrium is similar to a simple acid, note the difference in a titration curve, where the -log(H) is measured.

Electrochemistry 7

7.1 BIOCHEMISTS POTENTIAL

It is commonly known that physiological pH is around 7. But there is also a physiological redox potential, which seems to have somewhat more variability. A marker for the physiological redox potential is the ratio of reduced (GSH) and oxidized (GSSG), which can vary from around 20:1 to 1:1, and it apparently a function of health condition. Glutathione concentration is in the millimolar range.

Biochemistry commonly uses thermodynamic values at pH = 7, although the physical chemistry standard pH is at 1 M ion concentration. Thus, for an electrode reaction that includes a hydrogen ion, the potential will differ from the standard potential, E^0. Biochemists denote the potential at pH = 7 $E^{0'}$.

For a half reaction,

$$aA^+ + mH^+ + ne^- \rightleftharpoons bB$$

The Nernst equation is:

$$E^{0'} = E^0 - \frac{0.05916}{n} \frac{[B]^b}{[A]^a[H^+]^m}$$

$\frac{RT}{F} = 0.05916$ at standard conditions.

7.2 ASCORBIC ACID

Ascorbic Acid is itself an acid with K_a and K_2 and a $E^0 = 0.390$.

$$H_2A \rightleftharpoons A^{2-} + 2H^+$$

DOI: 10.1201/9781003358640-7

$$\text{Dehydroascorbate} + 2\text{H}^+ + 2e^- \rightleftharpoons \text{Ascorbicacid}$$

This system can be solved as follows [1]

$$E = E^0 - \frac{0.05916}{2} \cdot \log\frac{1}{[\text{H}^+]^2 + [\text{H}^+] \cdot K_{a_1} + K_{a_2} \cdot K_{a_2}} - \frac{0.05916}{2} \cdot \log\frac{\text{Red}}{\text{Ox}}$$

Assuming the reduced and oxidized forms are equal, we can calculate the potential as a function of pH.

```
k1 <- 10^(-4.1)
k2 <- 10^(-11.79)

pH  <-  seq(2,8,0.2)

H <-  10^(-pH)

Ecalc <-0.39 - 0.05916/2 * log10(1/(H^2 + H*k1 + k1*k2))

plot(pH,Ecalc,ylab="E", main="Ascorbic Acid, E vs pH")
```

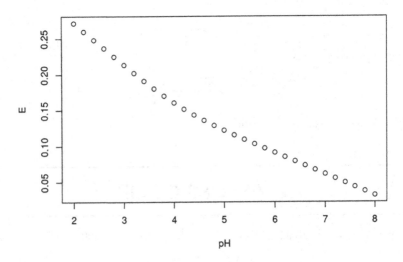

7.3 QUESTIONS AND PROJECTS

Consider the data from a classic PNAS paper [2]. Use nonlinear regression to find E^0 and K_{a_1} and K_{a_2}.

Table Potential vs pH, Ascorbic Acid [Red] = [Ox]

```
pH <- c(2.04,2.68,3.30,4.06,4.65,5.31,5.75)
```

```
Em <- c(.282,.242,.200,.163,.140,.117,.107)

Xdata <- data.frame(pH,Em)  # x and y to a data frame
knitr::kable(Xdata[,],
col.names = c('pH','E'),
caption = "Ascorbic Acid: E vs pH")    # makes a table
```

Table 7.1: Ascorbic Acid: E vs pH

pH	E
2.04	0.282
2.68	0.242
3.30	0.200
4.06	0.163
4.65	0.140
5.31	0.117
5.75	0.107

REFERENCES

1. D.C.Harris (2007). *Quantitative Chemical Analysis*, 7th edition. Freeman and Company: New York.
2. H. Borsook and G. Keighley (1933). Oxidation-reduction potential of ascorbic acid (vitamin C). *PNAS*, 19: 875–878.

Fourier Transform and Spectroscopy

8

8.1 AUDIO ANALOGY

Sound waves share useful similarities with electromagnetic radiation. They are both represented as traveling sine waves. Sound waves are a periodic disturbance in physical medium (air) whereas em radiation is a periodic change in an electric field. After a pulse excitation, the nmr signal decay contains that reflect multiple relaxations at different frequencies. To explore the nature of wave motion and the Fourier transform we can use audio files and R packages that create, read, and manipulate audio files (.wav).

packages:

fftw

TuneR

Below we used R to generate representations of sound. The graphs are of sound waves of different frequencies, as might be detected by a microphone: the plots are amplitude of sound versus time.

DOI: 10.1201/9781003358640-8

```
library(tuneR,fftw)

    Wobj1 <- 0.75*sine(880)
    # here is unit = freq not degree in radians
    tdir <- tempdir()
    tfile <- file.path(tdir, "myWave.wav")
    writeWave(Wobj1, filename = tfile)
    sdat1 <- tuneR::readWave(tfile)

    Wobj2 <- 1.0*sine(4400)
    tdir <- tempdir()
    tfile2 <- file.path(tdir, "myWave2.wav")
    writeWave(Wobj2, filename = tfile2)
    sdat2 <- tuneR::readWave(tfile2)


    Wobj3 <- 0.5*sine(2000)
    tdir <- tempdir()
    tfile3 <- file.path(tdir, "myWave3.wav")
    writeWave(Wobj3, filename = tfile3)
    sdat3 <- tuneR::readWave(tfile3)


    N <- 44100
```

```
x <- seq(1,44100,1)
w1 <- sdat1@left
w2 <- sdat2@left
w3 <- sdat3@left

wt <- w1 + w2 + w3

times <- x/44100
# same as freq because time = 1 sec

length(times)
```

```
## [1] 44100
```

```
par(mfrow=c(2,2))
# following plots 2x2

plot(times,w1,type="l",xlab="Time/sec",ylab="Amplitude",xlim=c(0,0.01))
plot(times,w2,type="l",xlab="Time/sec",ylab="Amplitude",xlim=c(0,0.01))
plot(times,w3,type="l",xlab="Time/sec",ylab="Amplitude",xlim=c(0,0.01))
plot(times,wt,type="l",xlab="Time/sec",ylab="Amplitude",xlim=c(0,0.01))
```

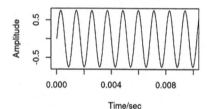
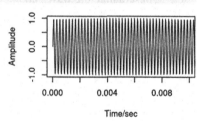
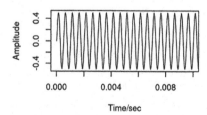
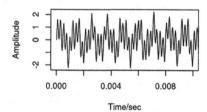

8.2 QUESTIONS FOR DISCUSSION

1. Consider the first graph. Imagine you are the microphone, and you measure amplitude by raising and lowering your hand.
2. According to the graph estimate how long does it take to go up and down once?
3. How many times will it go up and down in one second? This is the frequency.
4. Sound amplitude is related to compression of air. Sketch what compression of air might look like as a function of distance.
5. What is oscillating up and down in the case of EM radiation?

Now we will consider how the Fourier transform can translate a time domain representation to a frequency representation. For study of sound, and in spectroscopy, frequency representations are more useful. We use the R fftw package command FFT to do this. Now we have a graph of intensity versus frequency. We call this a transformation from the time domain to the frequency domain. For the individual time domain graphs, you could easily sketch a frequency domain representation. But what about the fourth graph? Somehow the FFT is able to unscramble the waves and do the conversion.

```
ft1 <- abs(fftw::FFT(w1))

ft2 <- abs(fftw::FFT(w2))

ft3 <- abs(fftw::FFT(w3))

ft4 <- abs(fftw::FFT(wt))

par(mfrow=c(2,2))

plot(x,ft1/N,type="l",xlim=c(0,5000))
plot(x,ft2/N,type="l",xlim=c(0,5000))
plot(x,ft3/N,type="l",xlim=c(0,5000))
plot(x,ft4/N,type="l",xlim=c(0,5000))
```

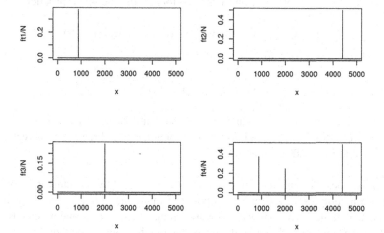

8.3 THE FOURIER TRANSFORM

As we have seen, the FT transforms time domain data to frequency domain. This turns out to be widely important and is key part of modern NMR (Earning Richard Ernst the Noble Prize), other spectroscopies (FTIR), signal processing, image processing, acoustics, and biology (bioacoustics).

Even though we may not encounter the Fourier transform in our math courses, we can try to get a conceptual understanding with the math we know. The key is in the idea of orthogonality.

8.4 QUESTIONS: ORTHOGONALITY

To try to understand what the FFT is doing, we can start with something familiar, a vector represented on x y coordinates.

1. Draw a vector $v(0.75, 0.50)$ on an x y grid. Now we know that if we drop a vertical line from the end of the vector, we will find the x component. How do we find the y component? Illustrate both.

2. In each case, more formally, what you just did was to multiply the vector $v(0.75, 0.50)$ by x and y unit vectors to find the scalar product.

$$\text{vec}_1 = v(1, 0) \quad \text{vec}_2 = v(0, 1) \quad \text{vec}_3 = v(2, 3)$$

Vector multiplication

$$v(x_1, y_1) \times v(x_2, y_2) = v(x_1 \times x_2, y_1 \times y_2)$$

3. What is the scalar product of the x and y unit vectors? The unit x and y vectors are orthogonal. We used orthogonality to break apart a vector into its x and y components.

Ok, what does this have to do with sound waves and the Fourier transform? We see that a complex waveform is composed of many individual waves. Now it turns out that the individual waves are orthogonal to each other, and we can use that fact to "pick out" the individual waves using the Fourier transform.

Since the math is more complicated, let's use R to explore.

If we multiply two functions and find the area under the curve of the resulting function, if it is zero then those are orthogonal.

We can sometimes see if the area under the curve is zero because the positive and negative areas are symmetrical.

using R:

1. graph the product for the following:

sin(x)*sin(x)
sine(x)*sin(2*x)
sin(x)*sin(5*x)
sin(x)*sin(2*x)*sin(5*x)*sin(x)

2. can use estimate the area under the curve of these graphs? finite or zero

The Fourier transform is simply doing this over and over again, with many frequencies. when it "hits" a part of the curve that matches the frequency (that is non-orthogonal) it has a non-zero result and picks that frequency out.

Here is an example of actual sound wave & fft using R.

```
# *******************************************
   # reading a music sample wav file


   mwave = "CantinaBand3.wav"
# loading the wav file


   data <- tuneR::readWave(mwave)


 # sampw = "sample.wav"


 #   sdata <- tuneR::readWave(sampw)


   summary(data)


##
## Wave Object
##   Number of Samples:      66150
##   Duration (seconds):     3
##   Samplingrate (Hertz):   22050
##   Channels (Mono/Stereo): Mono
##   PCM (integer format):   TRUE
##   Bit (8/16/24/32/64):    16
##
## Summary statistics for channel(s):
##
##      Min.   1st Qu.   Median    Mean   3rd Qu.      Max.
## -8587.000  -568.000    1.000  -2.556   588.000  9002.000
```

```
# sampling rate 44,100

ysig <- data@left

N <-  length(ysig)

x <- seq(1,N,1)

T <- 6.85

freq <- x/T

length(freq)
```

```
## [1] 66150
```

```
time <- T/x
ysig <- ysig / 2^(data@bit -1)
plot(time,ysig,xlim=c(0,0.002))
```

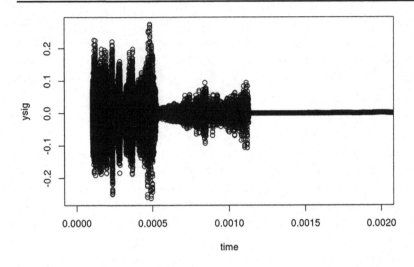

```
length(time)
```

```
## [1] 66150
```

```
par(mar=c(1,1,1,1))
# figure margin error fix


ffty <- fftw::FFT(ysig)      # discrete fourier transform

amp <- abs(ffty)
# DFT is a sequence of complex sinusoids

plot(freq,amp/N,type="l",xlim = c(0,2000))
```

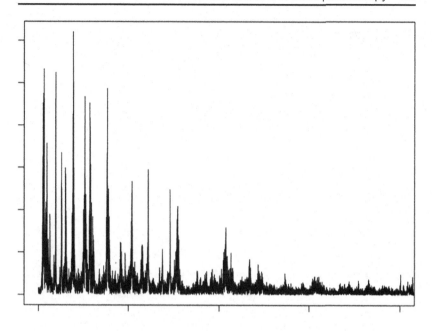

8.5 FT AND SIGNAL AVERAGING

Since FT spectroscopies are faster than continuous wave spectroscopies, more spectra can be collected, and we can take advantage of signal averaging. Use R to generate noisy sine waves and use signal averaging to see reduction in S/N ratio.

rnorm can be used to add noise to a signal. Here we simply demonstrate the impact of signal averaging by looking at the standard deviation as a function of number of signals.

after 1

after 10

after 100

How does the average noise change with n (=number of signals)

```
rn1 <- rnorm(10,1,0.5)

arn1 <-mean(rn1)

arn1
```

```
## [1] 0.7503965
```

```
rn2 <- rnorm(50,1,0.5)

arn2 <-mean(rn2)

arn2
```

```
## [1] 1.139269
```

```
rn3 <- rnorm(1000,1,0.5)

arn3 <-mean(rn3)

arn3
```

```
## [1] 1.00295
```

R Kinetic Analysis

9

Methods we have examined previously, such as titration or binding curves, have assumed a chemical equilibrium has been established. We now consider methods where a chemical system is moving toward equilibrium.

9.1 FIRST ORDER KINETICS

For a reaction

$$A \rightarrow B k_1$$

$$[A]_t = [A]_{t=0} e^{-k_1 \cdot t}$$

Below we simulate kinetic data with noise and the perform nonlinear regression with nsl2 In addition, we explore the use of a modified nelder-mead simplex, called subplex. Simplex methods are generally a bit slower than those used in nls2 (more iterations), but they are generally more robust. If you are getting an error using nls2 you can try subplex.

9.2 OPTIMIZATION METHODS: FIRST ORDER REACTION

```
# test of sublex (nelder mead variation)

# generate exponential data

t <- seq(0,9,1)

alpha <- 1.0 # Initial Concentration /M

k <- 0.1    # rate constant sec-1

rn <- rnorm(10,0,0.05)

y <- alpha*exp(-k*t) + rn

# First Order Concentration  + noise

plot(t,y,ylab= "M", main ="First Order Kinetic Simulation")
```

First Order Kinetic Simulation

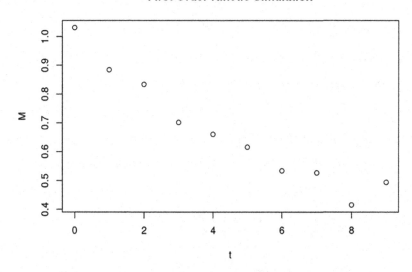

```
yfit <- 0

library(nls2)
```

```
## Loading required package: proto
```

```
library(subplex)

yfit <- nls2(y~alpha*exp(-k*t),start=c(alpha=2,k=0.5))

summary(yfit)      # nls or nls2
```

```
##
## Formula: y ~ alpha * exp(-k * t)
##
## Parameters:
##       Estimate Std. Error t value Pr(>|t|)
## alpha 0.994180   0.027569   36.06 3.83e-10 ***
## k     0.096598   0.006957   13.88 7.00e-07 ***
## ---
## Signif. codes:  0 '***' 0.001 '**' 0.01 '*' 0.05 '.' 0.1 ' ' 1
##
## Residual standard error: 0.04008 on 8 degrees of freedom
##
## Number of iterations to convergence: 5
## Achieved convergence tolerance: 8.122e-06
```

```
# exponential fit sing subplex

# y is previously simulated data set

# myfun is the sum of the squares of residuals

myfun <- function (x) {
  x1 <- x[1]
  x2 <- x[2]
  sum((y -  x1*exp(-x2*t))^2)   }

subfit <- subplex(par=c(2,0.5),fn=myfun)

subfit
```

```
## $par
## [1] 0.99418076 0.09659853
##
## $value
## [1] 0.01285161
##
## $counts
## [1] 375
##
## $convergence
## [1] 0
##
## $message
## [1] "success! tolerance satisfied"
##
## $hessian
## NULL
```

9.3 MONTE CARLO

Sublex itself does not provide error estimates on the determined parameters. However, since we are simulating experimental results if we simulate multiple experiments and find the average standard deviation it is consistent with results from nls.

The result of the average of five runs of sublex on simulated data results finds the following average parameter and sd.

Result of nls

alpha 1.02382 0.14057

k 0.12620 0.03813

Result from mean of five simulations with subplex
alpha 1.000769 sd 0.1478484
k 0.1050 sd 0.0519

The agreement is excellent, in view of the fact that multiple runs of nls on simulated data sets naturally result in slightly different values and standard deviations.

Subplex can be used in combination with a loop to mimic a Monte Carlo simulation as the loop allows for repeated iterations of a procedure, i.e. curve fitting, along with generated noise. Some common examples of curve fitting programs in R include nls and nls2, and while they're very good at what they do, there can be situations where they fail analytically and the Monte Carlo approach is more applicable. In the following example, this approach is used for a curve fit for the exponential decay of Iodine-131, with data obtained from the following site [1].

```
t<-seq(0,40,1)
A<-90*exp(-0.086427*t)
b<-c(0,0,0,0,0,0,0,0,0,0)
for (x in 1:10) {
  noise<-rnorm((t),mean=0,sd=1)
  myfun<- function(x){
    x1<-x[1]
    x2<-x[2]
    sum(((A+noise)-x1*exp(-x2*t))^2)
  }
  library(subplex)
  subfit<- subplex(par=c(80,0.05),fn=myfun)
  subfit[1]
  d<-subfit[1]
  b[[x]]<-d
}
v1<-unlist(b[1])
v2<-unlist(b[2])
v3<-unlist(b[3])
v4<-unlist(b[4])
v5<-unlist(b[5])
v6<-unlist(b[6])
v7<-unlist(b[7])
v8<-unlist(b[8])
v9<-unlist(b[9])
v10<-unlist(b[10])
initial<-c(v1[1],v2[1],v3[1],v4[1],v5[1],v6[1],v7[1],v8[1],v9[1],v10[1])
initialmean<-mean(initial)
initialstdev<-sd(initial)
initialstdev
```

```
## [1] 0.5323383
```

```
initialmean
```

```
## [1] 90.01724
```

```
v1<-unlist(b[1])
v2<-unlist(b[2])
v3<-unlist(b[3])
v4<-unlist(b[4])
v5<-unlist(b[5])
v6<-unlist(b[6])
v7<-unlist(b[7])
v8<-unlist(b[8])
v9<-unlist(b[9])
v10<-unlist(b[10])
kinitial<-c(v1[2],v2[2],v3[2],v4[2],v5[2],v6[2],v7[2],v8[2],v9[2],v10[2])
kinitialmean<-mean(kinitial)
kinitialstdev<-sd(kinitial)
kinitialstdev
```

```
## [1] 0.001093269
```

```
kinitialmean
```

```
## [1] 0.08649671
```

First, the raw data is simulated and a vector, b, was made for later in the program. Some noise is then generated using the rnorm command, and the standard deviation was found by using an nls fitting, but there are more sophisticated methods to find an appropriate amount of error. Then, the parameters for subplex to determine were defined with the exponential decay model; each iteration of the for loop would generate new noise for the original data and add the determined k-values and initial activity to the vector b. If more iterations were desired, b would have to be increased in length, but, while tedious, this is easily done. When subplex reports the findings for both parameters, it does so in a list format where both parameters are present together. As we want the parameters to be separated so we can investigate their respective means and standard deviations, the unlist command was used, and the desired calculations could then be done. The unlist command does exactly what it says: it unlists a list and turns it into a vector, which we already know how to manipulate from Chapter 1. When the iterations were complete, b contained information for each iteration, and each entry in b was then unlisted so the k-value information and the initial activity value could be added to different vectors. The mean and standard deviation of these vectors was then easily found, and the results were in good agreement with the researcher's determination of k and the value determined by nls.

9.4 SOLVING DIFFERENTIAL EQUATIONS WITH DESOLV PACKAGE

It is beyond the scope of this text to go into detail, but the R package provides a general differential equation problem solver (desolv) - a simple example is provided below.

```
library(deSolve)

chemkin <- function (Time, State, Pars) {
  with(as.list(c(State, Pars)), {
    dy =  -k*y
    return(list(c(dy)))
  })
}


Pars <- c(k = 0.1)   #  rate constant
State <- c( y = 1)    #  initial  value
Time <- seq(0, 10, by = 1)


#  don't need to stipulate data frame


out <- (ode(func = chemkin, y = State, parms = Pars, times = Time))


plot(out,ylab = "M",main="ODE Simulation")
```

ODE Simulation

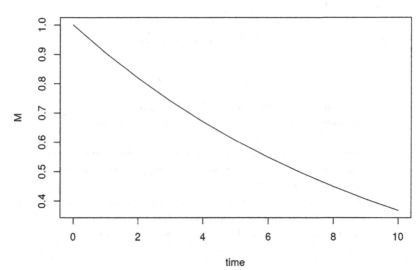

time

In addition, there is a package FME which couples with desolv to perform regression on data with the model determined with desolv. Examples are given in the package documentation.

9.5 ENZYME KINETICS

The simplest formulation of an enzyme reaction is:

$$E + S \rightleftharpoons [ES] \rightarrow E + P$$

where the first step is characterized by rates k_1 and k_2: and a slower second step k_3.

$$K_{eq} = \frac{k_1}{k_2}$$

It is generally accepted that enzymes catalyze reactions by binding the transition state and lowering the free energy of reaction.

Experimentally, studies of enzyme kinetics focus on initial rates. Using a steady state approximation - the concentration of ES quickly reaches a steady state value - the Michalis-Menten equation - properly derived by Haldane and Briggs,

$$\text{Rate} = \frac{d[P]}{dt} = k_3[ES] = \frac{\text{Rate}_{max}[S]}{K_m + [S]}$$

$$K_m = v_{max} = K_{cat}xE$$

an equation reminiscent of the binding equation.
Complications due to the progress of the reaction vanish:

1. inhibition by accumulated products, loss of activity of the enzyme
2. The reverse reaction can be ignored because it cannot occur until some products have had time to appear.
3. An initial-rate equation is much simpler to derive and use than an equation for the full time course of a reaction.
4. There is no drift in the pH or other conditions at zero time.

9.6 PROJECT

The kinetic parameters K_m and K_{cat} are provided for Carbonic Anhydrase. Analyze multiple simulations of initial rates versus substrate concentrations with subplex and determine the average and standard deviation of K_m and K_{cat}. Use an enzyme concentration 5 μM and a range of substrate from 0 to 50 mM.

Carbonic anhydrase

$$K_m = 2.6 \cdot 10^{-2}$$

$$K_{cat} = 4.0 \cdot 10^5$$

REFERENCE

1. https://radioactivity.eu.com/phenomenon/iodine_131.

Reports in R Markdown

10

RStudio Cloud contains a powerful document creation tool R Markdown, suitable for scientific reports, thesis and publication quality research papers. Some nice features

- Simple mark up language
- Latex style equations
- Sharing through rpubs or other systems
- embedded R code to generate graphs in document

Creating Simple Report with R Markdown
In rstudio cloud, we can choose file → new file →
A new file (.rmd) will be created which already includes the first part of an rmd file the "YAML" header, which declares the title, and html and pdf outputs.

""markdown—
title: "Sample Document"
author: "D. Gosser"
date: '2023-06-12'
output:
html_document: default

The default created R Markdown file also includes examples of simple formatting and embedding r code.

Now you can input and format text, equations, images, and references, and R code. You can view your formatted document: Select "knit" to produce an html or pdf document.

Headers and subheaders are created with #, ##, ###.

""markdown Header Levels

DOI: 10.1201/9781003358640-10

Header 1
Header 2
Header 3

"" Below are examples of R Markdown code and the output they produce.

10.1 BASIC FORMATTING

```
**Boldface**
```

Boldface

```
[I'm an inline-style link](https://www.google.com)
```

I'm an inline-style link

10.2 EMBEDDING R CODE IN R MARKDOWN

```r
x <- seq(1,20,1)
y <- x^2
plot(x,y)
```

```
<img src="_main_files/figure-html/unnamed-chunk-37-1.png" width="672" />
```

```r
x <- seq(1,20,1)
y <- x^2
plot(x,y)
```

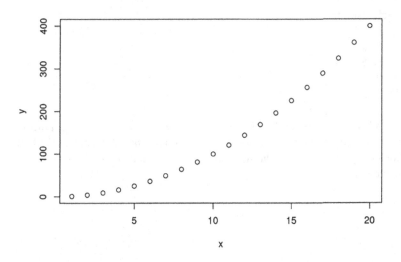

For scientific work, the ability to insert latex equations is very useful. To insert an inline equation bracket the latex code with dollar signs.

Examples of latex equations

y^2

k_{cat}

$a \cdot b$

$\frac {a \cdot b}{c}$

$\alpha \cdot \beta$

$$\alpha \cdot \beta$$

which result in:

y^2

K_{cat}

$a \cdot b$

$\frac{a \cdot b}{c}$

$\alpha \cdot \beta$

$\alpha \cdot \beta$ (next line, centered)

After knitting to html, a simple way to share is to publish the document in an rpub account, which provides an attractive html based document. In Rstudiocloud, after you knit to html, choose "publish" and choose rpub and create an account.

Index

Printed in the United States
by Baker & Taylor Publisher Services